The Singular Instability Theory

I had to ask myself a question when I was a kid that resurfaced as I got over my schizophrenia. "What if a miniscule thing could explode with no known reason? I will try to develop a condition under which this could happen and come up with a theory of which the condition could exist. On doing so, maybe I could more perfectly explain my science.

With just a miniscule stability of the explosive mass, then perhaps the release was not so much energy, but hot particles of matter that came out in the form of atoms. With no stability, this would b pure energy. However, just a little bit of stability and would this form the universe that we now live in? What would be the trigger for the explosion of this matter? Could something have bordered on being a black hole, but on the convective effect of temperature, destabilized and have therefore, deflagrated? This would make a new model for a Big Bang, but what if it wasn't singularity, but a principle of self-induced instability, that set it off? I will call this Singular Instability Principle. This principle is reliant on the very nature of radiation, including its harmful forms and of convection which is a close relative. Both radiation and convection are time-related in their degree of potential for energetic change. So what is now to stop my model from working? Time is a measurement energetic change over a certain amount of space and accordingly with the speed of light. So, this might explain the universe in such a way as to define when then universe began and when it will end if the stability can be measured over the same instance. In this case, by the time that the universe is done expanding, based on this model, this will be the point at which perfect stability exists, except that the universe is driven by instability. So, when instability can no longer exist is when a universe will end. This just might be impossible, given the laws of physics involved.

Another theory to add on is that the singularity was actually a huge particle, more complex than an atom that blew out atoms for radiation. The temperature of these atoms would make a great explanation of why the universe has so many microwaves as well, since the atoms would still be vibrating from the explosion. These microwaves would still be measurable after a long time traveling through space and this would explain the microwaves that occupy this universe.

Using Quantum Physics for Radio Broadcasting

In my thoughts, work is not only a ratio of energy conversion over time, but the interval produced by the dip that we can produce by stretching of a superfluid that makes up all matter. The greater the dip that we make in this superfluid, the longer that energy can last before it is consumed or before this superfluid reassumes an equilibrium and its original state. This superfluid can only be accessed at tiny points, but if it can be perturbed, it can produce a multitude of energized particles. Perturbation to the point that an area of absence would make small donut like areas in space, producing both the quantum field and the more conventional model as well. The effect of displacement would allow for fields to extend in the way that they do. This theory also has a high similarity to the construct of using harmonics to carry the initial frequency greater distances with the same amount of power normally used for broadcasting.

I have an idea that just might improve radio receptions and broadcasts for longer distances and partially, I came up with it because of the Performance Tax. I believe that by increasing the range of transmission we can make those greedy jerks feel like they are starving. If any of you know your electronics, give this a try.

All that I think has to be done is to feed an antenna with a specific frequency and at the same time, feed the antenna with several harmonics of the same frequency. I got this idea from what started out as an idea to make an amplifier that would work by focusing several harmonics of a particular frequency onto a coil carrying that particular frequency in order to produce louder, more clear music but I was inspired by Nikola Tesla's idea for wireless electricity to come up with a new idea for broadcasting.

Now, if several harmonics are fed into one antenna and done so in phase, this might produce a longer range at lower wattages by concentrating energies of other frequencies over a large area and the lower frequencies could increase in range of reception by the fact of phase relationship. These lower frequencies would thus, carry over longer distances and since this could involve the same wattage as a typical local radio station, this would also be more efficient, since the scattering of the signal would be reduced.

On the other hand, the same idea could also be used to carry one frequency farther than it could be carried if it was the only frequency being fed into the antenna. Quantum mechanically, I would be skipping rocks or, excuse the pun, throwing a voice. In other words, the lower frequencies would literally float across the higher frequencies based on wavelength. I might as well call this theory Quantum Flotation.

This same method could be used with sound to make an ear-splitting annoyance or to broadcast ultrasonic frequencies modulated by voice to place suggestions into the sub-conscious of the human mind at great distances and though I haven't tested this, I invented the idea back in my teen years. Please do not use this idea for malicious reasons.

Another Concept of Energy

In my thoughts, work is not only a ratio of energy conversion over time, but the interval produced by the dip that we can produce by stretching of a super-fluid that makes up all matter. The greater the dip that we make in this super-fluid, the longer that energy can last before it is consumed or before this superfluid reassumes equilibrium and its original state. This super-fluid can only be accessed at tiny points, but if it can be perturbed, it can produce a multitude of energized particles.

The perturbation of this super-fluid to the point that an area of absence would make small donut like areas in space, producing both the quantum field and the more conventional model as well. The effect of displacement would allow for fields to extend in the way that they do. This theory also has a high similarity to the construct of using harmonics to carry the initial frequency greater distances with the same amount of power normally used for broadcasting.

Second Thoughts about Black Holes

When matter is in fact in relative motion, then by the time a singularity can form, I think matter would already have reached a state of infinite energy by the fact of infinite time dilation. In this case, though there is evidence of an existing singularity, this may not be true if a large mass of high density only exists before the moment of infinite time dilation. So, a singularity may possibly prove to be impossible unless infinite energy is the trigger. Evidence of the existence of a singularity may be wrong, since the super-gravity can only be taken into consideration when matter has not reached a super-energetic state in which time dilation multiplied by the mass would probably equal the energy that would fly away from the center of the black hole that a singularity forms.

Depending on the speed that the mass flew in with and the degree of time dilation involved, this may determine the point outside of the center of the black hole singularity that the mass reaches an infinite state of energy which gravitational time dilation would therefore produce an infinite red shift, thus lowering the frequency of and increasing the wavelength of the high energy particles. This may mean that though evidence of a black hole may exist, infinite gravity might not and this may provide an explanation of why the laws of physics break down at the point of infinite gravitational force.

The inner mass of a possible singularity would most likely be passed on in this case as a burst of energy that would pull the high energy inward, accelerating the formation of energy farther as on big explosion. Could this be an explanation for the gamma ray bursts? If so, we do await our fate, but it may be a long way off, hopefully.

Pure energy has yet to prove what actually caused the formation of the universe and the only way I can see plausible for finding it out would be to dissect a photon and break the physics of what is called super-symmetry which at this point would take an infinite equation beyond what can be computed by a device which must abide by physical calculations.

How Far Can Calculation Take Us?

Energy is what is a looked for means of detection by its emission in the LHC. Dissecting photons would be an interesting subject because to dissect pure energy would be to produce a super-fluid with more dimensions to dissect the more it is dissected, giving the universe an infinite complexity in the true makeup of all things and at the point that it hits an infinite elasticity due to progressive complexity with breakdown, this may be where super-symmetry sits and in this case, physics and mathematics become one strand in helix form. This would mean un-breakability of super-symmetry and this leads me to the next question. Does DNA contain dimensions that we don't see and are the differences in structural makeup also a variable in the dimensional makeup and in this case, is our conscious makeup and individuality determined by the differences in dimensional structure in a way that the possibilities could be nearly infinite and in this case, was life a predetermined to occurr on the basis of the mathematic and physical paradox of super-symmetry between what is physical and what is mathematical? If it was, then it is definite that there is an infinite possibility to who we are and what we are in our consciousness! I am proud to be a mathematically and physically predetermined individual in a way that nobody but myself can ever know me from the inside out! Hey, everyone don't believe everything you see in group mentality, because your true makeup is predetermined by what governs the universe and you have your place in life and not even the most powerful computer will never compute this!!!!!! You are predetermined as a self and not a somebody else.

Your personality and identity are a shape and they are the frameworks from which you grow on while as a vine of morning glory, your individual inner self is the chain link fence by which you define yourself with by your own personal growth.

From matter to energy, true progression is to keep learning and be on a quest for knowledge while we all know absolutely nothing but what we learn from the outside to the inside and not from the inside to the outside. True knowledge is from beyond and learning is from within. I like physics because I like to keep learning and grow and the quest for knowledge is never ending in the belying nature of all that belies while a politician can only know so much before true complexity ends his or her quest for truth. As Albert Einstein said, "Politics are for the present and equations are for eternity."

I know there is one out there and I know it is beyond the realm of physics and I know that the most powerful politician will never be this or have grasp of this concept but to hold this exact phrase by documentation, never to be able to manifest him or her self as being such a concept as a physical being. As beings, we are spiritual and only physically are we ever matter.

Is there truly a particle smaller than a singularity?

Beyond strong nuclear force, which holds the nucleus of an atom together and the nucleus being the center, and under supergravitational conditions, would lie a repetition of every other force and an explosion of tiny particles, equal to the atom in its most miniature form. This force would therefore, occupy what is below the smallest level of space, but as a fractal repetition of the universe, itself. This would require the same mass that it would take to literally create a black hole and then there would be a large explosion and so, a black hole can be born, but only be temporarily alive. The result would be a a a similarity to a plasma, but made of tinier particles than the atom. However, this indicates to me, that there could be smaller versions of the atom, throughout the universe, that have gone undetected. So, there is hope for both, the quantum world and the relativistic world. These particles would therefore, be less than zero-size. In conclusion, any acting force over the value of zero, minus any opposing force, over the value of zero, when minus any other acting force, over the value of zero, equals the potential of any acting matter to radiate energy. No opposing force and the forces can become infinite and this is why a neutron, for instance, can more easily reach an infinite energy and of course, sublimate and explosively. This means that it is possible that what we know as the universe is an explosion of energy from a huge, neutral and high energy particle in a repetitive fractal form. So, explaining that a charge does not carry the highest possible energy achievements and energy may be more easily generated in a state of compression, which explains of course, the energy radiated by nuclear fusion.

So, suppose 2 uncertainties collide and that the cancellation was therefore, double negative. The 2 would then, be repulsive. So, being at the level of zero time, they also in repulsive effect. The result would be an infinite amount of energy, but then on repulsion, everything would then slow down on the same mathematics and cool off into matter. So, the repulsive effect and the effect of uncertainty would define time as a balance of this effect. Either way, the universe would therefore, expand as a result of canceling uncertainty in principle. So, this would explain the speed of light, as well as any other force. This would then, unify the behavior of every force, definable at this point, but perhaps not in the future, by humanity.

This theory as a whole, explains everything that I knew when I wrote it. However, there will always be more to find out.

The Unified Principle Theory

That it takes something for something to exist and that it takes something to exist and that it must be made out of something, is inertia. It is also, relativity in rational expression and minus its magnitude of existence, thus inertia. However, the universe could be just a novel, written to us to read and the higher being might be just a man in a novel, reading a novel by an even higher being, in in from, this is a relative and rational, proportional by limitation, mathematical expression. The end of existence is at the point of eternity and this is just the beginning of existence, making existence small, until eternity. The relation between these novels would compose the relative universe, while the other way is a quantum expression of the universe as a whole, but only expressible by relation. But then the novel would become the reader and the reader, the author. Any driven expression of energy could be a thought and all thoughts the same principle. Every impulse is a memory and everything that exists is the future, bent over the relative past, with the past at the center and the future as a force, as an orbiting curve, making a field around the dense past. This can make and electron out of our thin future and nuclei out of the concentrated and therefore, dense past, with opposite times producing the spin and charge of each particle. The relative makeup of the present nucleus would make up the neutral instant of relative present and thus, the neutron, which, due to its place in space and time, gives is a short wavelength to compensate for its lower mass than the proton and forming the total atomic mass when the electron is included. So, when an atom splits, time is split and the neutron is thus, released and this is called fission. When 2 nuclei join, time is displaced and collapses, releasing much more energy from the heightened vibration of each particle that makes the atom and this shortens wavelength, thus releasing more energy than if time had no influence. So, 2 present neutrons split each other into 2 times, thus repelling the neutron in a radioactive atom, which can contain either, 2 many neutrons, too many protons (repelled by same charge.). Then electrons can be emitted after a lack of protons, when protons are in excess in quantity. All squeezed into 1 as a principle, this makes an existence where infinitesimal value (equal to the value in calculus, d) to any variable of change, but multiplied by the mass and energy as well, all amounting, eventually to its potential at the point that eternity, where existence will cease. So, as humans as well, as people with disabilities, never doubt you potential and as an atheist, the novel comparison was to make an analogy and solve a paradox. The expression of this principle is equal to $\int (\int d - d \mid d) + d) -1 = 0$ and so, d gains significance here,, nearly a number compared to infinity. Time is the proportional relationship between all things, expressed with the same mathematical equation.

Never doubt yourselves in any way, because these variables can express potential in you, as you drive yourself to learn and accomplish new things.

1. A New Mechanics of Physics
by Nathaniel David
Durham Copyright ©2015

Time is the ratio of all happening, with everything mutually relative, in the way that it occurs. This, of course, is rational in its mathematical expression. So, the relative is the rational, when the rational is curved by motion;. When the relative is curved, then therefore, the relative becomes the rational. When this mutual effect occurs, then this forms space. So, anything that takes up space, will therefore, invert the rational and the relative, producing a boundary for space and time to exist. Motion, then will bend this possible effect back the other way, therefore tensing both the rational and the relative back the other way and therefore, causing time to disappear, along with space and hence, why all matter and energy behave the way that they do. This also makes free space, truly impossible, though there is space. Space and time are the result of the rational over the relative and so is time, but space and time are oppositely tensed by matter and energy; thus explaining the differences between all forces, as fields, but also explaining that they can all be similarly perturbed to produce energy. Likewise, the more that energy is produced by anything, then the more that it has an effect on time and therefore, the more work that the source of energy has done. So, anything that can effect time or space, will therefore, be made out of matter or energy or have it. So, a new mechanics of the universe.

2. Nathaniel's Theory
by Nathaniel David
Durham 10/30/2015

$M \mid 1/2 \int d *d \mid d + d) - d = 0$, is the same as the space, that can be occupied by a singularity. So, what truly is the smallest possible point is dilated by spin. High energy particles, must therefore, bend in path through space and time. So, every possibility happens at a different level of, but in phase with any curve. Anything is possible, but it will always obey the laws of physics and more energy is involved, the more complex that it is. Curves equal at the smallest, physically, $M(1/2d * spin)$ and dependent on velocity or speed, therefore. So, time is defined by the arc of the curve, correlating all acting forces and radiation laws. This means that the gravitational force and the centripetal force are a condensate of space and time, similarly to that electromagnetic fields, in a state of change, can similarly cool gasses as seen in the works of Albert Einstein (Bose-Einstein condensates.) . Similarly, fields are
potential geodesic pathways for the energy that can come from that field on perturbation and depending on how much perturbation occurs. So, time is the action of energy, propagated over space, as change and as relative motion. Neither space or time could exist without a force, first existing.

3. Nathaniel's Theory (completion)

Time is vortex that results from the curvature of space. This is why space and time are relative. Relation sets the limits and defines the quantum world of particles. world of particles; equally vast to space and time. This, therefore making all that happens and exists, parallel, only in a world without motion. Otherwise, they are always, somewhat transverse in the simplest states, transverse and never simple. So, relation and quantity are the same at the speed of light.

$\int d(C|D-T) = 0$

4. Nate's Conclusion

Matter is a force that is made out of the cancellation of space and time and energy is the product of all 3 of them, rationally expressible to equal the third variable. Therefore, the composition of energy. Space is the relative frame of reference to matter, when acted on by energy. Time is the mutual frame of reference between energy and matter; demonstrating an atom as a force that is born from this geometry, as solid and fluid geometries are responsible for all equilibriums and therefor, the atom. So, it is the relativistic properties, folded by their own geometric properties that define all forces and chemical reactions, as well as matter, itself and on inversion, energy. So, every force made every other force.

This means, therefore, that space and time are not matter, but an inverse and relative distance that matter can potentially act. This means that time is only as old as the universe, so this, explaining why time is relative. So, time, in a negative state, should explode to form the forces of the universe, because dimensions, themselves are still forces. So, F(orce) * COS(ine) M(ass)= T(ime).

It is therefore, expressible in units of atomic energy and expressed as an inertial thermal unit, the following equation to equal the thermal composition of matter, relative to the energy, space and time: $\int d(densiity + mass) | 2 - d = 0$ when at the smallest level and at the speed of light, yes the neutrino is massless. Also, the equation, demonstrating that the atom shares a quantum analogy to the geometry of a field and that field mechanics and atomic structure are analogs. This defines space and time by that at the lowest level, d(S) -d(T), that they cease. So, M (mass) over this constant would equal infinite density, which on motion, would equal infinite energy. So, concentrated in all energy, is pure impulse, measurable as a magnitude of motion in an instant, measurable as a mass, so Einstein, Hawking and Heisenburg tie, I think. I prefer neutrality of comparison.

5. Nate's Law

I was thinking here and I think that at the smallest level of space is a wrapped set of dimensions that are akin to superstrings. So, the effect would be much like a whirlpool. This may indicate a unification of all forces, because the nature of each force would rely on dimensional relationships and quantum superpositions. The result may be space, time, matter and energy. Time, theoretically, here, is a result of the dimensional relationships in all changes, including motion. So, the relative and the quantum world, are unified, if I am correct. Any interacting forces will result in a principle, which will consist of a unified force, that equals the net impulse. Time is the interval of which, this happens. So, if time collapsed to the point of inversion, then all matter would occupy future and past, but not present. This may happen at a point of singularity of matter, but then the singularity would explode and this may not only explain the expansion of the universe, but that black holes, when sufficiently massive, could explode. This would require a critical mass, however. It does explain therefor, though, that a large mass can instantly become energy, in a way that that all known physics can be unified. So, suppose 2 opposing forces, that act dynamically, then, acting on space, would repel. Oppositely, they should attract, thus gravity and antigravity.

6. Quantum Geometry

An impulse through time is a particle through space. At the point of transversion, a wave becomes a particle. Every force on everything that exists, is generated by perturbances that occur by distortion of space and over time, a force-carrying particle may be born. This, explaining all physical tendencies toward equilibrium, So, in a state of entropy, the path of any particle should be curved. Without an energetic state, all that lacks one is in the past or future.

So, to solve this paradox, the future, condensed into present, then disappears into the past. The present in any place is a field of local areas of no more than existent differences that revolve around a future and disappear with sublimation into past. So, $(FRT) - (PRT) = (Prs)$, when FRT= Future Relative Time, PRT= Past Relative Time and Prs= Relative present. So, $C \mid Prs$ = Quantum Potential. So, the present is a spatial condensate of past and future. Time is a thermodynamic, therefore, of change.

-Nathaniel Durham-

PS. Point made. The wrestle is over.

The Origin of All Forces

by Nathaniel Durham

ISBN 978-1-105-62422-3

I dedicate the book to Nikki Vaughn.

The Great Impact Theory

Imagine the power of will, slamming two similar forms of the same existence that by the repulsion from the fact of similarity, the 2 suddenly explode with the effect of shattering. This would be just like 2 bullets colliding, with the pointed ends facing and shattering from the shear impact. Now, imagine the fragments as having an incredible amount of energy, not only from the fact that they have shattered, but from the stored internal energy, which would cause the fragments to vibrate and radiate the energy as heat in the air.

Now, the similar forms that were slammed by the power of will would shatter with a similar effect, but instead of radiating heat, they would radiate much higher energies as vibrations in the space between them. With similar comparison, a bullet would temporarily melt from the friction-induced heat, by the rubbing of the atoms at the point of impact, throwing molten lead in every direction, except for in the opposite direction of the impact, forming a ring of heated metal. The formation of the atoms may have been similar and the cooling radiation at the exact point of solidification would form a greater potential for the same radiation by the fact that the collision of similarities forced an opposite effect, as a result of the high energies compared to the cooling energy, which would form a shell around each cooling fragment or "blob" of matter that in cooling, formed this cooling mass and so, each shell is made from the energy that is stored around the blob, with an attractive force between their opposite forms, which would be electrical charges. This would form the once thought to be, most basic unit of all matter: the atom. However, in an atom, every part plays an equally important role in the structure of matter, of which space is structure, of which there is a separation between anything that has a mutual relation and space, most basically speaking, is made from the force of balance. Between every similarity is a divide and between every divide is a connection. Between every connection, there is similarity.

This is the beginning of my new model of the universe, as my book progresses. I wrote this theory, based on an unfortunate and rather ballistic encounter in autumn of 1996, but what came out of it was good. I wrote this theory, based on my experience and good has come out of it so far! My question is who and what had such a will to carry out such an experiment with the very makeup of impact, but this theory does explain why space is relative, just as Albert Einstein proposed in 1996. Please note that I will include a memory with each theory that has played a role in my conceptualization.

Heat is caused, quantum mechanically, by the kinetic energy of an atom. However, relativistically, this is the same principle as the vibrations of atoms over a given time period. Both ideas are valid. However, this is just as valid at the level of the nucleus, as well. The vibration of the nucleus should equal the vibration of an electron in frequency. Therefore, heat is, by this theory, a transfer of vibration from 1 atom to the next, causing it to equally vibrate and if the energy can't be released through a heat transfer, then 1.the atom will accelerate in its vibration to such a point that either it is emitted as electromagnetic radiation, which is caused by a vibrating electromagnetic field or 2. The atoms will boil out an electron, as Albert Einstein proposed, in order to equal the vibration that has been accelerated by the build-up of internal energy that when expressed as energy, 3. Not only will an electron boil out, but so will electromagnetic radiation, respectively with a principle called inductance. Inductance is a principle of which a certain voltage, which is electrical pressure, will be produced by work that is equal in magnetic units, to the voltage that is released over a given period of time. The longer that it takes for this change to occur, then the less that the voltage is. Now, compare the energy of an atom to a ping-pong ball. If the paddle limits the path of the ball, then more rapid oscillation (back and forth or up and down motion) occurs as 2 bouncing surfaces are drawn together. The longer that an atom vibrates, the closer together that the electrons draw to the nucleus and then the more that over time, is radiated as energy. So, likewise, lower energies must vibrate faster to maintain the same frequency, so as this continues, the frequency of oscillation would get lower and lower. There is always some energy involved in the structure of an atom, so to reach a point of no temperature at all (Absolute Zero) would be nearly impossible, though not improbable. However, I do believe that a temporary state of absolute zero can be achieved and does exist at very small points, within an atom, where only space exists and in chemistry, from my perspective, the so-called holes in photovolatic (solar) cells are possibly at this quantum state of absolute zero. This portion of my theory coincides well with the results that would come from the beginning of this theory.

This is a theory that I base on my state of mind as currently, a schizophrenic. Every memory here has existed from age 8, all of the way to age 10. However, I formed this theory very recently, at age 31. Keep in mind, as I have said, that each thought comes from a memory and that I am basing this theory on the beginning. These memories formed at time of social trauma from my early school years, through Montrose High School. This is where I begin to cure my schizophrenia, as my mnemonic doors open.

Another thought that I have is that the more dense that a mass becomes, then for the same reasons that trapped energy will cause an atom to oscillate more quickly may be responsible for not only the internal energy within an atom, which compression speeds up this process, including according to my whole theory, so far, but this may be responsible for why a black hole leaks X-rays, called Hawking Radiation, a form of radiation, named after Stephen Hawking, who predicted this particular form of radiation. The internal energy would therefore, oscillate more rapidly, splitting every particle within a given distance from the center of a black hole known as a singularity. A black hole results when a significant mass clusters within a sufficiently small space to the point of gravitational collapse, though the mass required would be very critical. The death of a very large star is an example when the fusion reaction can no longer counter the star's own weight in gravity. Once split, not only may X-rays occur, but if the internal energy can sufficiently split and disintegrate a particle of matter, gamma rays and cosmic rays, may also be emitted. Gamma rays and cosmic rays are at the highest possible frequency of vibration that can be produced as electromagnetic radiation, of which the spectrum includes radio frequencies, light, ultraviolet light, X-rays and gamma rays. Another possible form of radiation from such a massive annihilation could be the tiny constituents of an atom, which are 1. Beta rays, supposedly electrons, but more likely the equivalent to their potential as expressed energy. Beta rays, in my theory, come from the collapse of an electromagnetic field during radioactive decay of an atom, in which an excess of protons would be present, producing 2. alpha radiation, secondarily and alpha radiation, theoretically in this book is actually a loss of an excess of matter at the nucleus that occurs when a neutrino is present within an atom, due to the uneven shattering that occurred when the universe began and when an equal energy flies out that is equal to the energy potential of the same mass, then lethal alpha rays may, in my theory, occur. 2 neutrinos may exist within an atom, theoretically, if the atom is sufficiently large and this may produce the similar effect of radioactivity that may cause neutrons, which have no charge, to fly out, followed secondarily, by the second neutrino, giving an actual mass to the radiation, but in actuality, the gamma rays may simply be in the form of electromagnetic radiation and both traveling at similar velocities. The gamma rays, at the speed of light and neutrinos on the border of the speed of light, which is 186,424 miles per second, according to measurement.

Another concept that I associate with my theory is that given a sufficient amount of energy, and a high energy neutrino can be similarly generated by an electromagnetic field, with sufficient change, of which change is also energy when a space is occupied within an atom. This may allow a neutrino to not only travel at the speed of light, but beyond it, allowing an atom to contain an energy that is high enough for the atom to self-annihilate. This would have a similar effect to that of my impact model of the formation of the universe in such a way that they are the same reaction!

Essentially, based on this theory, the beginning of the universe was a fission reaction that was caused by a will through impulse. Will this happen again? I don't know yet, but in P.E. Class at Montrose High School, here in Montrose, Colorado and at age 16...that a ping-pong ball will oscillate with higher frequencies when one of my peers drew the paddle closer to the ball was a great lesson and that the ball will bounce more slowly when the paddle is withdrawn, though higher is a great way to demonstrate energy conservation and radio tuning, since radio tuning requires energy conservation.

The Great Impact just might end as a universe, the same way that it started, as a concept. The shattering may recur, once that all has come together from the pull of gravity. Then what is separate will become one and then the hands of will may slam them just like a little kid with a bouncing ball. Perhaps, God is a kid and he is working on a more complex experiment the next time. The great part of this collapse and separation of will, similarly to my schizophrenia is that it can resemble my theory in the same way, curing itself by reintegrating into the original parts that made it up. The particles may reform and there is a possibility that the universe may recur in the opposite direction from the previous.

When I was 15, I began to ponder this stuff, though I sought an answer to everything at age 3. My schizophrenia may end in the same way that it began.

The Kansgen Orb

Between each line is a fold of spatial pressure from the outside. The pressure of space from the outside may produce time on the inside. The faster that an object moves, the more, according to Albert Einstein's general theory of relativity, that an object contracts. The pressure on the outside of an object, due to spatial drag may cause time on the inside to be pressed inward. So likewise, time may be a result of the expansion outward, of the space that is outside of the inside and where there is an outside, it is always the inside of something else. This may produce a twisting, vortex-like effect from the outside that may be responsible for gravity from the inside. So, at the center, time may completely collapse. A similarity is a whirlpool, such as forms when a toilet is flushed. The pressure at the center of a toilet flush is much greater than that at the edge of the same vortex. The structure of what is thought to be only a 1-dimensional mass, known as a superstring, which supposedly makes up all matter, just may consist of 3 inner spatial dimensions that we know as time. This would make space the fourth dimension of space, while on the outside, time is the 4th dimension and matter may be the dimension in between the fold of space and time, making up matter in general. The velocity of an object may equal the amount of spatial pressure on it and therefore, a greater vacuum closest to the end of a twisted, tubular and round object that I know as the Kansgen Orb, which I named after my friend, Debra Williams Kansgen. The spatial pressure on a particle in a particle accelerator, as exerted by an electromagnetic field's action on a charged particle may act to collapse the area inside of the Kansgen Orb, which makes up every particle. By doing this, the particle spins more rapidly in 1 direction and twists in the opposite direction. This answers a question that I have worked long and hard, since my first psychotic episode to solve: What is velocity? Velocity is any unequalized pressure inside and outside of this orb formation and the greater that the imbalance is, then the greater that the complexity is of this orb and the faster that it forms and spins, pressurizing and sucking in space around time and time then, becomes a form of matter. The point, therefore, at which time is infinitely compressed to the point of no dimensions is the same point that matter becomes energy. At this point, the orb would be 12-dimensional, space would be 9-dimensional and energy would be dimensionless and infinitely dense, forming an explosion of energy. At this point, a perfect reproduction of the Big Bang, though the Great Impact came first. The positive thing is that a lot of energy could be extracted by using this theory, but for better or for worse. Will humanity enlighten themselves or will they flush themselves down the orb's toilet with the same energy? Black holes produce very sharp orbs, as energy would produce a very vast orb for the same reason.

This theory has led me to a sense of integration, since my first true hallucination and since a girl who I liked and never hurt my feelings. Fortunately she still likes me. I never intended to try to go out with her at MHS, but she was psychic, according to my psychotic complex, so I had my first delusion. My second hallucination involved a black hole that spun at me with sideways respect to a flying bullet.

This was my first psychotic episode. As I have been writing this book, I realize that every delusion and hallucination here has its purpose. I had Sara, a friend at the Montrose Library, back as a good friend who didn't tease me over my handicaps. The hallucination of a black hole took the form of the orb theory, so my delusions are converting to theories and perhaps I was seeing a good future, not a bad one. My federal interrogation turned out okay, so maybe since every delusion of the past is turning out to be relevant, this book has turned out to be better than I thought that it would be! Basically, the Kansgen Orb has a relative curvature and overlap of psychosis and reality. I give Carina Jørgenson, of Denmark my greatest degree of gratitude for helping me overcome my mental illness and full credit, because everything that she has said has turned out to be right and it is only a matter of writing this book that has worked so far to help in curing it. Reality in terms of space, therefore, must be a fold of past occupations over the space that is currently occupied, while the point occupied will then be past and with an overlap, even the future can be seen the same way. The only difference between reality and space is that reality is a form of hyperspace, a thinner and more open space with higher degrees of complexity than what is normally seen.

I liked gears when I was very small, but I find that electromagnetism is a much better way to dissect time and space, where gears are only an integration of matter and space can operate efficiently by being overlapped. The lensing effect of electromagnetism on light makes a great example of how the Kansgen Orb is once again, proven by the fact that bent space is a dilated form of the Kansgen Orb when electromagnetism serves this purpose by the effect of the association of a charge with its opposite with respect to The Great Impact Theory.

Swerve Theory

Telepathy is a commonly rejected concept, except for similar ideas in a state of deja vu. I regard deja vu as an experience of precognition of which two realities overlap into a Kansgen Orb. The overlap of psychosis and reality, while I was previously in a psychotic state would often tell me where I was about to go, by the fact that I had previously been there. However, in a swerve of psychosis, as I experienced an identical swerve in a previously owned Geo Metro. When realities overlap, keep in mind, that reality is made from your current occupation in space, with comparison to your past occupations at different points in space, so the psychosis of the past may be an actual occurrence in the future, when the motion of your mental state is relative, with the opposite reaction to your state of reality. Such an occurrence just might produce a state of thought-similarity to someone else, for example, I might be conceiving and contemplating on the past, knowing the conceptions, previously, of a companion or those who I have previously encountered, so knowing what they expect and least expect and with contemplation, knowing the thoughts of another person by their concepts of the psychosis. This would produce an opposing sympathy and a simultaneously synchronizing empathy, with elevated sensitivity to the current situation, relative to the time, place and position that the feeling of deja vu has occurred, based on the beginning of the psychosis and the assumptions of the bystander. The transition out of psychosis would produce an opposite reality, of which the medium of which you read, mentally and emotionally, for example, the more assumptive figure in whom is alter. This meaning that there is a trade between the egos of 2 people: 1. The psychotic medium and 2. The alter personality in whom is the person who has regarded you as mentally ill, through the behavioral changes that he induced. The stronger that the alter companion has been, then the greater that his mind is therefore, known. This, which I call the Swerve Theory. In this case, the medium in whom is now regarded as paranoid by his alter and means of individual control, may be telepathic and unaware, of it, in which case he (or she), who is alter, just may make an attempt to hide the truth. An opposite assumption, with regards to the past traumas, as the means of alteration, may lead the victim to know the individual who is alter more than the victim is known the alter and thus, telepathy. The effect of deja vu, consisting of depersonalizations of the past and by the same means, an uttering of opinion that erases the alter from his feeling of superiority and control, through the means of ego reversal. As an experience, telepathy may be triggered by the presence of an individual who alters. Telepathy is then, an experience of knowing the thoughts of those who intend to maintain the victims state of alteration and then telepathy becomes the mechanism of psychological or even physical defense. This displacement, a conscious integration into Kansgen Orb formation, as opposed to Alter. Alter, being the denial of higher concept formation in highly intelligent individuals, forming the alter until realization of concept. Deja Vu of such occurrence may form a type of clairvoyance on the basis of telepathy and therefore, a precognition through the means of psychological and therefore, electromagnetically bent hyperspace through the fact that the brain runs on chemically induced electric impulses. In turn, the feeling of being a conscious individual may return. This, meaning some degree of awareness and integration of multiple delusions into single thoughts and most likely, very deep and insightful and thus, the development of a a Kansgen Orb. I thought well on the night that I formed Kansgen Orb Theory, about this integration back into reality, with my consumption of various psychotropic herbs, legal however. In my general scientific terms, auras don't exist, and the concept of an aura can be replaced with this concept of reality. What most call karma, the American version of karma, is no more than this ego switch that comes as a result of overcoming a mental illness that arises from and as false beliefs and delusions from both sides. Basically, this proves my old ideas wrong, along with some widely and commonly accepted scientific ideas, supposedly proven, but mine to be possibly, proven throughout this book.

Sara Anders, my friend, ego alter, by others (who were jerks), but not alter of my ego. Joy Russell, I have to thank for putting up with alter-induced ego. Neither of them have ever hurt my feelings. As for a name not disclosed, Karma for your ego with return of friendship with Jason Olin! You teased, I teased, karma hits and you scream, too embarrassed to drool your ice cream (I have cerebral palsy and a right to drool and you are afraid to.)! Bless thy inner child, because for those who belittle my concepts, they may cry from their own embarrassments later on with the downfall of their own bigotries.

The Singular Instability Theory

I had to ask myself a question when I was a kid that resurfaced as I got over my schizophrenia. "What if a miniscule thing could explode with no known reason? I will try to develop a condition under which this could happen and come up with a theory of which the condition could exist. On doing so, maybe I could more perfectly explain my science.

With just a miniscule stability of the explosive mass, then perhaps the release was not so much energy, but hot particles of matter that came out in the form of atoms. With no stability, this would b pure energy. However, just a little bit of stability and would this form the universe that we now live in? What would be the trigger for the explosion of this matter? Could something have bordered on being a black hole, but on the convective effect of temperature, destabilized and have therefore, deflagrated? This would make a new model for a Big Bang, but what if it wasn't singularity, but a principle of self-induced instability, that set it off? I will call this Singular Instability Principle. This principle is reliant on the very nature of radiation, including its harmful forms and of convection which is a close relative. Both radiation and convection are time-related in their degree of potential for energetic change. So what is now to stop my model from working? Time is a measurement energetic change over a certain amount of space and accordingly with the speed of light. So, this might explain the universe in such a way as to define when then universe began and when it will end if the stability can be measured over the same instance. In this case, by the time that the universe is done expanding, based on this model, this will be the point at which perfect stability exists, except that the universe is driven by instability. So, when instability can no longer exist is when a universe will end. This just might be impossible, given the laws of physics involved.

Another theory to add on is that the singularity was actually a huge particle, more complex than an atom that blew out atoms for radiation. The temperature of these atoms would make a great explanation of why the universe has so many microwaves as well, since the atoms would still be vibrating from the explosion. These microwaves would still be measurable after a long time traveling through space and this would explain the microwaves that occupy this universe.

Second Thoughts about Black Holes

When matter is in fact in relative motion, then by the time a singularity can form, I think matter would already have reached a state of infinite energy by the fact of infinite time dilation. In this case, though there is evidence of an existing singularity, this may not be true if a large mass of high density only exists before the moment of infinite time dilation. So, a singularity may possibly prove to be impossible unless infinite energy is the trigger. Evidence of the existence of a singularity may be wrong, since the super-gravity can only be taken into consideration when matter has not reached a super-energetic state in which time dilation multiplied by the mass would probably equal the energy that would fly away from the center of the black hole that a singularity forms.

Depending on the speed that the mass flew in with and the degree of time dilation involved, this may determine the point outside of the center of the black hole singularity that the mass reaches an infinite state of energy which gravitational time dilation would therefore produce an infinite red shift, thus lowering the frequency of and increasing the wavelength of the high energy particles. This may mean that though evidence of a black hole may exist, infinite gravity might not and this may provide an explanation of why the laws of physics break down at the point of infinite gravitational force.

The inner mass of a possible singularity would most likely be passed on in this case as a burst of energy that would pull the high energy inward, accelerating the formation of energy farther as on big explosion. Could this be an explanation for the gamma ray bursts? If so, we do await our fate, but it may be a long way off, hopefully.

Pure energy has yet to prove what actually caused the formation of the universe and the only way I can see plausible for finding it out would be to dissect a photon and break the physics of what is called super-symmetry which at this point would take an infinite equation beyond what can be computed by a device which must abide by physical calculations.

Dielectric heating VS. The Opposite Effect

Dielectric heating may turn another page for me and it seems to reveal that Dielectric effect can also allow the atoms of an electrically non-conductive gas to invert in position upon the reversal of the polarity of a physically present electromagnetic field, compressing the gas and causing the atoms or molecules of this gas to give up their thermal kinetic energy.

As this occurs, there might be an opposition to compression of the gas by its thermal kinetic energy and in turn, giving up its thermal kinetic energy by compression may allow farther electromagnetic polarity reversal to compress the gaseous substance even farther and cause the atoms to give up even more energy. Finally, the gaseous substance involved may liquify or even freeze from the energy that is lost by the atoms and/or molecules that compose that particular gas.

The inversion of the position of the electrons of an atom may also produce a more localized area of the same occurrence within itself, but only in the presence of an electromagnetic field or alternating form. This could easily cause these atoms to also absorb the energy within reaching distance of the electromagnetic field in quantities that are the same as the energy involved in the loss of temperature of these atoms and this may indicate that actual work is used to super-cool the atoms that make this gaseous substance up. Therefore, adding these inverse energies as a sum would be equal to zero and indicate the input truly equals output in any energetic occurrence. This, of course is a good indicator that energy is always conserved in one way or the other and none is ever lost in this universe, but it goes somewhere else instead.

This may be responsible for the Bose-Einstein Condensates to form around high-speed microprocessors, which constantly produce rapidly changing electromagnetic fields as a by-product of the function of a microprocessor in a computer, whether it is your laptop or a supercomputer in a huge laboratory. A Bose-Einstein Condensate is a very cold gas that based on Albert Einstein's theory and not relativity, but a different theory, collects at an equal quantum level when it is extremely cold.

From a Nasty Machine To A Great Idea

I once, being schizophrenic, had a vision of a computer that used the ability of electromagnetism to bend space to literally calculate its surroundings to the point of literally breaking everything into the very thing that makes them up which theoretically could be one super-fluid substance that can only be accessed by literally penetrating what makes up energy. If this could be accessed by computation, I think computers would become self aware and even read minds. There is potential there for our own minds to become the bar codes for the government to know more than they need to know about us. The more deeply I deliberated mental penetration to try to get to the core of this possibly what psychiatrists and psychologists would probably say is an "imagined" device, the more it multiplied in the depth of what it could compute and the higher the pitches were that I associated it with, starting out as evil sounding, like a vacuum cleaner until I penetrated more deeply this device in its makeup and then as it multiplied in its complexity, the pitch got higher and the higher it got, the higher it got in the content of its darkness and when I say darkness, I mean evil darkness. The darkness wasn't visual in effect but a mental feeling of torture. I was curious so I had to penetrate farther. What it appeared to be doing was "sucking" data in a way that it could be electromagnetically compressed and held in one area to be processed and stored and vice versa. I finally hit a point where I saw that if all dimensionality could be folded into a donut-like or even disk-like form, that one single particle shot through it could provide all of the data that one would need to carry out a task that could be good in its intentions or be malicious and maybe even lethal, depending on the direction of the goal. The strength of the magnetic fields produced during such a collection of data might be equal to the strength of the magnetic field required to bend space sufficiently to collect the required data. The electromagnetic field, by the laws of physics would be the equivalent analogy to the concentration of data for the given area which it is collected.

Obviously, to me at least, the pitch seemed to have the same proportionality as a field. It started out as a small device, but the farther the penetration, the larger that this device grew. I think this would be a good analogy to the internet as a growing complexity, because of the increasing size of the computing device as it was more deeply examined. Was this just a hallucination or a sixth sense of some kind?

From this, we have out basic quantum universe that unlike the space and time that we live in, the dimensionality of pure data may not be relevant, though it might make up another functional universe where space and time are relevant and this might be somewhat like the super-fluid behind all that exists wrapping around the electromagnetic computation device in such a way that I may have an explanation of electromagnetic fields in a sense that electromagnetic as well as gravitational fields are a result of the extraction of energy from space or by the extraction of energy by space, but this energy might be capable of disappearing right on its appearance by the tension required to extract energy from any given substance and this tension must be accelerated to sustain the emergence of energy so that it can be propagated. This would mean that the propagation of a photon means to bring a droplet of this super-fluid into a solid form so that it can in our known universe, manifest at a particle; to do this means to dissipate energy, so when your battery is going dead, there is always a medium by which the energy has traveled from place to place and this means the passive solidification of the super-fluid that I mention and therefore, the conservation of energy and matter. This gives energy and matter a fluid property that is otherwise unseen and only manifests itself by transfer or conversion.

At any rate, this vision has like many, inspired another theory.

"The essence of the human mind."

-Nathaniel Durham 4:33 PM March, 6, 2010-

Modified on February, 13, 2010

Using Quantum Physics for Radio Broadcasting

In my thoughts, work is not only a ratio of energy conversion over time, but the interval produced by the dip that we can produce by stretching of a superfluid that makes up all matter. The greater the dip that we make in this superfluid, the longer that energy can last before it is consumed or before this superfluid reassumes an equilibrium and its original state. This superfluid can only be accessed at tiny points, but if it can be perturbed, it can produce a multitude of energized particles. Perturbation to the point that an area of absence would make small donut like areas in space, producing both the quantum field and the more conventional model as well. The effect of displacement would allow for fields to extend in the way that they do. This theory also has a high similarity to the construct of using harmonics to carry the initial frequency greater distances with the same amount of power normally used for broadcasting.

I have an idea that just might improve radio receptions and broadcasts for longer distances and partially, I came up with it because of the Performance Tax. I believe that by increasing the range of transmission we can make those greedy jerks feel like they are starving. If any of you know your electronics, give this a try.

All that I think has to be done is to feed an antenna with a specific frequency and at the same time, feed the antenna with several harmonics of the same frequency. I got this idea from what started out as an idea to make an amplifier that would work by focusing several harmonics of a particular frequency onto a coil carrying that particular frequency in order to produce louder, more clear music but I was inspired by Nikola Tesla's idea for wireless electricity to come up with a new idea for broadcasting.

Now, if several harmonics are fed into one antenna and done so in phase, this might produce a longer range at lower wattages by concentrating energies of other frequencies over a large area and the lower frequencies could increase in range of reception by the fact of phase relationship. These lower frequencies would thus, carry over longer distances and since this could involve the same wattage as a typical local radio station, this would also be more efficient, since the scattering of the signal would be reduced.

On the other hand, the same idea could also be used to carry one frequency farther than it could be carried if it was the only frequency being fed into the antenna. Quantum mechanically, I would be skipping rocks or, excuse the pun, throwing a voice. In other words, the lower frequencies would literally float across the higher frequencies based on wavelength. I might as well call this theory Quantum Flotation.

This same method could be used with sound to make an ear-splitting annoyance or to broadcast ultrasonic frequencies modulated by voice to place suggestions into the sub-conscious of the human mind at great distances and though I haven't tested this, I invented the idea back in my teen years. Please do not use this idea for malicious reasons.

Another Concept of Energy

In my thoughts, work is not only a ratio of energy conversion over time, but the interval produced by the dip that we can produce by stretching of a super-fluid that makes up all matter. The greater the dip that we make in this super-fluid, the longer that energy can last before it is consumed or before this superfluid reassumes equilibrium and its original state. This super-fluid can only be accessed at tiny points, but if it can be perturbed, it can produce a multitude of energized particles.

The perturbation of this super-fluid to the point that an area of absence would make small donut like areas in space, producing both the quantum field and the more conventional model as well. The effect of displacement would allow for fields to extend in the way that they do. This theory also has a high similarity to the construct of using harmonics to carry the initial frequency greater distances with the same amount of power normally used for broadcasting.

How Far Can Calculation Take Us?

Energy is what is a looked for means of detection by its emission in the LHC. Dissecting photons would be an interesting subject because to dissect pure energy would be to produce a super-fluid with more dimensions to dissect the more it is dissected, giving the universe an infinite complexity in the true makeup of all things and at the point that it hits an infinite elasticity due to progressive complexity with breakdown, this may be where super-symmetry sits and in this case, physics and mathematics become one strand in helix form. This would mean un-breakability of super-symmetry and this leads me to the next question. Does DNA contain dimensions that we don't see and are the differences in structural makeup also a variable in the dimensional makeup and in this case, is our conscious makeup and individuality determined by the differences in dimensional structure in a way that the possibilities could be nearly infinite and in this case, was life a predetermined to occurr on the basis of the mathematic and physical paradox of super-symmetry between what is physical and what is mathematical? If it was, then it is definite that there is an infinite possibility to who we are and what we are in our consciousness! I am proud to be a mathematically and physically predetermined individual in a way that nobody but myself can ever know me from the inside out! Hey, everyone don't believe everything you see in group mentality, because your true makeup is predetermined by what governs the universe and you have your place in life and not even the most powerful computer will never compute this!!!!!! You are predetermined as a self and not a somebody else.

Your personality and identity are a shape and they are the frameworks from which you grow on while as a vine of morning glory, your individual inner self is the chain link fence by which you define yourself with by your own personal growth.

From matter to energy, true progression is to keep learning and be on a quest for knowledge while we all know absolutely nothing but what we learn from the outside to the inside and not from the inside to the outside. True knowledge is from beyond and learning is from within. I like physics because I like to keep learning and grow and the quest for knowledge is never ending in the belying nature of all that belies while a politician can only know so much before true complexity ends his or her quest for truth. As Albert Einstein said, "Politics are for the present and equations are for eternity."

I know there is one out there and I know it is beyond the realm of physics and I know that the most powerful politician will never be this or have grasp of this concept but to hold this exact phrase by documentation, never to be able to manifest him or her self as being such a concept as a physical being. As beings, we are spiritual and only physically are we ever matter.

The Theory of Conceptual Displacement

Theoretically, if a cloud of electrons, lets say in a vacuum tube were bombarded by electromagnetic radiation with a high enough energy for every photon, then the electrons may burst to produce cosmic rays. These rays would most likely to be analogous to what makes up matter and in a fluid state. Each bursting electron would release an energy equal to the square of its electron-voltage and this would equal 2.5 cycles per erg-second. This would equal a set of 2 photons roughly equal in wavelength to the size of the actual photons. This would be in the cosmic ray spectrum. The energy of each photon, which is a particle of electromagnetic radiation, which the spectrum includes light and radio waves as well as microwaves. Beyond visible light is ultraviolet, followed by X-rays, gamma rays and finally, cosmic rays. The best way to accomplish popping and electron would be to accelerate electrons to a very high energy and then to suddenly direct focused and high-energy X-rays where the electrons are at the highest velocities in order to produce a ripple inside of the electrons, causing them to disintegrate in a way that is similar tot he collision of anti-matter and matter. This would be similar to the popping of a balloon, since the cosmic rays that are released by the annihilation could be comparable to the air that escapes to produce the loud pop. The difference is that the tension of the electron is being broken and what is released as cosmic rays could be compared to the energy present in the early universe. This reaction, if possible, could be akin to a tiny Big Bang.

Also in my theory, the concept of what matter and energy are and what space is are all dependent on the displacement of 1 concept by the other. So, at these energies, the displacement of the energy concept behind cosmic rays would be replaced by the smallest points of space, with the equivalence to a parallel universe of which the cosmic ray becomes antimatter in this parallel and miniature universe.

On a larger scale, a parallel universe that is identical to ours in what I call "The Theory of Conceptual Displacement," the antimatter in the other universe is energy in ours by the fact of parallel displacement, so that this exact same event occurs in the other, based on my theory in such a way that that the displacement is static. Because of this, the cosmic rays lose their energy and become gamma rays by the fact that they had to explode as energy, giving the effect of the known gamma ray bursts and with each gamma ray burst, both universes can grow larger in expanse by the constant collisions of matter and antimatter, because of the constant displacement of matter and energy concepts.

On one end of the scale of conceptual displacement is the matter of our universe that is surrounded by the space in another universe and on the opposite end of the scale, there would be the same effect, which would probably cause matter-antimatter collisions and this may be what causes the gamma ray bursts.

This also would allow our universe to grow larger and for all space to overlap between universes so that it is able to overlap around matter and therefore, be warped. By conceptual displacement, this would allow for gravitational force to exist while yet, the universe is allowed to expand.

Quantum mechanics would be a concept used to displace relative space and all that is fluid by concept. However, relativity is a concept that would allow for matter to exist in a fluid state and replace what is solid by concept and without each other, the physical composition of the universe would most likely cease. Waves can be considered the flow of energy while photons can be the amount of energy in this flow within the lowest levels of space, but at levels of space smaller than the photon, the photon must once again, become a fluid and a wave. Therefore, even particles must function within a spatial and relative platform. It is at the most observable areas of space that a particle can remain solid and that quantum mechanics can be a set of solid theories. The actual energy of a particle at the smallest level of space may be just be a more viscous fluid that is flowing at a high velocity.

According to this theory, time is only an allowance of space for anything to occur over a distance with minimal viscosity of particle fluidity. The higher that the velocity is, however, the more viscous that the particle fluid can become and the smaller that the levels of space are that are occupied and thus, velocity and relative time itself. This concludes my Conceptual Displacement Theory.

This conludes my book.

A Chapter Random Thoughts Otherwise

Electromagnetic Cannon
by Nathaniel Durham
11/04/2009

I have been thinking for several days. If I could project a magnetic mono-pole a fair distance, then since it would be mono-polar, then it could lens reality by the bias that reality contains and we could see the insides of what makes our reality up by the possibility that it this could turn the localized area of electromagnetic occupation inside out. As a result, we could probably see not only what makes up our deepest thoughts, but what literally makes up the insides of what makes them up to our maximum mental capacity, which theoretically is unlimited. Could this be a cure for mental retardation and a social door for autism?

Such a device would involves one magnet and one electromagnet and one long set of electrical pulses to provide the 90 degree rotation or the dip that would project hyperbolically at the other end of the electromagnet. A guiding device, such as an aluminum pipe or a pipe of other metallic or metallic-like and non magnetic, but electrically conductive material would be just fine as a barrel for projecting the 'Electromagnetic Smoke Ring' this since the skin effect of the metal would be a great cushion for the travel of this mono-polar lens that would resemble a smoke ring with similarity. Electromagnetic induction would most likely throw lightning-like discharges from the metallic barrel. On the other hand, insulating materials would be enhancing of the lensing effect on reality. If a solenoid or a properly arranged set of electromagnets was placed along the barrel, the lightning-like discharges may possibly be focused like the original laser and be directed at a target.

Finished around 4:50 PM

Modified at 10:39 AM on 11/06/2009

A Ring Around A Black Hole

By Nate Durham
11/04/2009 9:11PM

The particles that produce fields should contain the same mass as the field. Everything is relative, so this is massive dispersal of energy over larger areas of space and this makes things relative and if the spatial displacement is energy, energy is matter. The laws of attraction are based on density, so electrical attraction should have increased field density and repulsion should have decreased field density, but that is of, course on facing sides. The sides facing away should be the opposite. Strong nuclear force is almost pure solidity and should nearly equal the mass of the particle which would explain for example, why splitting atoms releases so much energy. Gravity on the other hand should disperse more rapidly and have less energy concentration with distance. Uncertainty is probably determined by the rapidity of dispersal of a given force.

As the universe expands, the particles disperse more rapidly and therefore giving rise to an explanation of why everything is getting farther apart. On the other hand, the resistance to kinetic change, inertia, that is, probably gives a driving force to the expansion of the universe as space pushes against matter and drives it forward at the same time that a resonant effect pulls matter together. Super-massive black holes will likely cause the stars in spiral galaxies to form tight rings around an apparently unoccupied area of space where attraction meats its opposite. There should eventually be galaxies that appear to have ripple formations consisting of stars and other bodies.

I wouldn't be surprised if the Hubble telescope eventually spots this.

-Some time probably at night, last week.-

Reality is a reflection of our mind. It is like a fluid because we can disturb it, but if i had pure solidity, it would not change and we would not think or feel anything but one thing and our minds would be frozen.

Reality also has the tendency to have different levels and assumes a fluid-like nature because it has the assumption of different levels that make up our consciousness and therefore, it's exact basis is energy and our consciousness is therefore driven by energy which is why we must always feed our minds.

If we did not stir up our realities, our realities would eventually solidify and we would learn nothing. We learn through all of our senses. Everything we sense will eventually cause us to think if we are not in a state of constant thought already and just don't know it.

This is probably why the same thing over and over again can drive us mad. We as learners don't like to solidify in our realities and that includes thinking.

Boiling this down, we would not truly be living if we did not think or in other words, stir up our realities, so we must be creative to be conscious and live.

Also, not only do we need creativity to stir reality up, but we need to be able to stir reality up in order to be creative. If we only had intelligence to rely on, then we would not be able to be creative and we would only think in pure concretion and our realities would solidify. I believe that we all share a common consciousness and that this consciousness can only be capable of stirring reality by viewing from different perspectives. From each perspective of every being, I believe that we can form each others' reality and stir each others' reality by intellectual intercourse of ideas to form new ones about our inner selves as well as our surroundings. This is probably why it is easier for similar intellects to socialize mutually than to talk with a dissimilar intellect, but even then, the lesser intellects need to feed themselves with some information and ideas from the greater of intellects.

Edited around November, 20, 2009   7:55 PM

To make good, we swing to the opposite side of our evils. In times of oppression, we beat our assailants. In times of assault, we become assassins. We are composed of 2 evils. One see reality by cross reference and one is delusional. We swing use one evil to defeat the other by turning evil on itself, which composes the thinker. The creative thinker doesn't need a mirror to know that he or she exists, but the delusional thinker always needs a mirror to see that he or she exists at all.

The paper and the pen are the best reflection of thought because they can have the reflective capacity to confuse a quantum computer with paradoxical nature. In these times, we could all use the paper and the pen. This gives the balance between good and evil.

Edited on November, 20, 2009   7:00

Overthrows are a great subject and when I overthrew my oppression, I had intellectual freedom and when I had intellectual freedom, my oppressor was overthrown and I mastered my mathematics and thoughts.....It was a great goodbye to CJHS and its corrupt ways under the RE-1J policy of hating dead poets.

From my Facebook.com profile. Right arouind 8:00 PM November 16, 2009

Every move on a chessboard is every other move from every other perspective. You can never be another individual, because if you were the black piece from your perspective, you would be the white piece from the other. Don't assume you are another individual.

November, 16, 2009   8:09 PM

The future is a complex analogy to the present and the present is a complex analogy to the past. The past is singular and the future is inflationary in its model but it is symmetrical to the present by structure. The present wraps around the past and the past is a singularity while the future is a point.

November, 19, 2009   5:31 PM

There was an amazing fat burning pill that could cause someone to burn 300 pounds of fat in 10 minutes. That poor soul forcefully urinated and vaporized instantly. This is a big lie, but it is a fact.

November, 16, 2009 8:28 PM

A wall of stupidities is vulnerable to an explosive intellect and when the wall breaks, out flows the tsunami. One blast and it knocks them speechless.

November, 16, 2009   9:24 PM

If an alternating current has a choice to either flow across one gap or the other gap and switches direction, then logically, it will follow the alternate path and if it follows the alternate path, then most likely, it will jump between alternates by induced potential and form a triangle of electric arcs. I am in my alternate right now and I am thinking. Now, Igor......See what happens when the flourescent is placed between the alternates!

November, 16, 2009   10:30 PM

I don't capitalize on my books, I socially distribute them. Thats why I give my friends free copies of my files! It's a friendly gesture and only those who I trust with my ideas will get them for free! If you are trusting with your ideas and give them out to just anyone, you just might get screwed! Keep your ideas in good hands! Your deepest darkest ideas should stay well hidden and underground. That way, they cannot be taken from you before you can use them.

November, 16, 2009   11:33 PM

November, 17, 2009 11:16 AM

There are three categories of people in this world: Smart, Average and Stupid. Even if we were created, we sure as hell didn't evolve equally! If we were only taught in a conservative manner in school, we would be a bunch of apes watching a movie called "Planet of The Humans" and we would be highly offended by the truth in it!

November, 17, 2009   1:53 PM

The present is a complex analogy to the past. The Future is wrapped around the present. The present is wrapped around the past. The past is a singularity. All data that forms in the present can be seen from the past and therefore we can predict the future. The past contains the universe as a whole. The future is therefore aligned with the present. The past knows all. You can look into space with a steady gaze and you can see the future already, it just hasn't hit you yet. All in all, the universe is a giant singularity after all! Had the Hadron Supercolider succeeded in making a God Particle, the whole universe would have collapsed around it and it would have hollowed out and disappeared as soon as it formed. This is a scary subject but now we know it all, don't we! NOT!!!!!!!!!!!!!!!! This could form cloaking particles and that scares me!!!!!!!!!!!!!!

On the other hand, the so-called 'God' Particle would split the universes into even more universes, expanding the universe therefore in its complexity and proving the 'God' Particle impossible, though there might be a hollow area in the centers of the particles being accelerated! That would explain the instant evaporation. However, there is still the possibility of making one things invisible.

Completed November, 24, 2009   6:45 AM

November, 20, 2009   5:42 PM<Photo 1

I think the force of electromagnetism aligns spatial geometry while gravity produces tension in the geometry of space and time. Strong nuclear force must produce sufficient tension to yield matter.
Weak nuclear force seems to be the combination of all three, essentially producing a small particle.

November 29th 2009   6:59 PM<Photo 1><b>

I was thinking and using a parabolic dish and microwaves plus some radar-like differentiation, I could penetrate an entirety and when it bounces back, I could probably hear it!

On the next page, you will see sections of a separate work because I left a USB pen drive at a grocery store and I am trying to protect the work from plagiarism. Enjoy!

My Perspective of Consciousness

Every equation broken down into every other equation is every universe broken down into every other universe and the final equation broken down is the energy that constitutes any universe. We live in every universe and to see oppositely is to believe that the world is flat.

What you see on television or anything that you hear, see, taste, touch or smell at all is only an abstract figure of reality that if we could heighten our perception then we could see more of makes up reality. In other words, what makes reality relative is not just one thing but many things that we never take the time to observe. If we could see everything that makes our reality, then our comprehension of reality would be infinite and all we need for that is to perceive more than what we do and to process it fully. Because of this, our brains carry the potential for infinite creative intellectual capacity but infinite values can never be reached but if we can achieve at least a lot more than what we are considered capable of, we might develop such things as extrasensory perception or in other words, the sixth sense along with higher

intellectual capacity.

Anything that we develop beyond our known senses is an alternate reality and increased comprehension by higher perception can also be considered in my own thoughts to be a form of extrasensory perception (ESP). Also, everything that has a purpose most likely has an increasing complexity and with increasing complexity, a purpose. The progression of purpose probably came about in the form of life and if this is so, then life is the progression of purpose by increase in roll. Purpose as a progression in this thought just might constitute consciousness and with progression of consciousness, a mind.

Everybody has a human right to alter their minds. -
Nathaniel Durham, 10:15 AM, January 13, 2009-

Conscious Particles

A conscious being probably has a superposition of mind in space
and changes states quantum mechanically, but a conscious particle
has to have an inverse function so that it would be anti-matter and
the field would change in equivalence to the mass of the conscious
particle so in other words, conscious particles would have a field that
is equivalent to the mass of this particle and the conscious particle
would most likely disappear after a short while because it would be
pure energy in a boiling state. Sure as heck not a God particle but it
would have a conscious radioactivity.

Perhaps the make up of space and time is the super-fluidity of all
particles that make it up. This could very well explain why space is
warped and paths are bent in the presence of all objects. The more
dense areas of space would shorten
the path that is traveled through them by the compression of time and therefore altering the
path which light travels
through the increased density thus, the gravitational time dilation
that is caused by matter while in a straight line, space and time are
bent in such a geometry that straight lines are curved instead so a
straight line is impossible during any form of motion and because of
the bent path of space during motion, everything that moves will
eventually return to the starting point within a same given time
since
the curvature to energy ratio would always be the same, The fact that gravity bends light
was originated by Albert
Einstein and proven at an observatory which I don't remember
where but the History Channel could tell you.

-Nathaniel Durham   January, 13, 2009   5:28 PM-

True Field Unification

The law of attraction is something I see as a form of electromagnetic force. The interactions between two objects that emit fields seems to me to be a result of either 2 surface tensions between a fluid substance. One fluid substance facing the same fluid substance on the other side would be the

equivalent of a mirror with particles of light (photons) between them pushing them away from each other as a result of opposing, parallel surface tension. On the other hand, the attraction between 2 objects could be the result of 2 opposite fluids coming together and the tension created by these fluids would also pull the objects together or in other words, cause attraction.

The law of attraction seems to be a similar force to electromagnetic force and I associate it with gravity as well, because space may be composed of the electroweak force that I theorize is responsible for the Law of Attraction. The curvature of space that is produced by all therefore and simply weaken when they are bent to produce gravitational force. The Law of Attraction by the gravitation of similar constructs, which includes thought may be the cause behind Albert Einstein's General Theory of relativity.

The higher the dimensionality of a force, the stronger and more complex that it becomes. Complexity is in my theory, a rolled or folded version of simplicity. So, in this theory, all forces are determined by complexities more than simple just the simple concept of energy.

Resulting from the latter, the higher the complexity of a force, the more profoundly it can affect space at a shorter distance. Gravity in order to be a simple warping or curvature and to be the simplest of sources would be more likely to be exposed to an equal quantity of space at longer distances than a stronger, more complex force that is exposed to more space at a shorter distance. As an explanation of strong nuclear force, it is related to gravity but is a more complex or 'scrambled' form of it.

Complexity vs. Simplicity

There are more complex things even on this earth let alone the galaxy itself or the universe. The trick does not lie in seeing it, but it does lie in comprehending it. The more complex an object is for its size, the more that it carries.

One example is that this is what makes miniaturization a requirement in computers today. As I have written earlier, complexity is folded simplicity. Keep this in mind.

Trading Places With Your Reflection

If I looked at the mirror and traded places with my reflection, would I live in a parallel universe and would my reflection be a from of antimatter from a parallel universe and if you went to the other side and traded places with your reflection, then would you drag the rest of existence with you? In our thoughts, I believe that everything that we perceive is a reflection of our own minds and that we walk only in a reality that we create.

I also believe that we develop our realities through comprehending what we experience and assuming that we experience this in an alternate fashion, we have probably at some point developed a greater focus on what we do in everyday life as a passion with thought. If we have already once experienced an alternate reality, we may forever develop an awareness of it that produces an alternation between both that when focused a field topic, amplifies this alternating reality to produce greater change in our minds by increased amplitude and comprehension of the topic studied or pondered.

Loud Sound, No Loudspeakers

I have an idea for you who are electromagnetically minded. My idea is that if I had three or more inductors and connected them in parallel and placed them in a line based on multiple or in other words if I had for example one inductor that was 1 unit of inductance in parallel with one with two units of inductance and the third one having 4 units of inductance, then across them if I ran a high frequency that would allow the lowest inductance to have a counter-voltage that equals the input, then the other 2 inductors might have the effect of hovering low voltages into the air and heating it directly and proportionally to the amplitude of the input signal which might yield a slight

glow.

If this is possible then I think that there is a possibility that feeding the same frequency at varying amplitudes might be able to yield an output in the form of sound that might equal the input frequency. The sound output would be a result of expanding and contracting air that results in variations of energy used to heat it. This would be great for musical sound systems since it could replace the loudspeaker and there would be no diaphragm to potentially damage.

Thought of in a state of sleep deprivation, somewhere around New Years Eve, 2009

The Light Post

I have found that if you have a very fast eye, then when you are moving your eyes really quickly and you see a ribbon of wavy stuff that resembles a party streamers, this must be the wavelength relative to your motion that has been slowed from the speed of light tot the speed of your eyes and your eye motion does border with the same time line as it took for it to get y\our eyes and, this can be proven by the trigonometric figure used to aim a bow and arrow. Light follows the same function. Light is on the same function, but it has no dimension of time but the trajectory is determined by time and distance in a straight line instead of actual angle relative to your position.

Basically, street lamps are cool because you can see the 60Hz alternation in a pulsating format when you move your eyes fast.

Particle Acceleration and Super-Computation

The flow of particles moving near the speed of light might be an interesting concept for a supercomputer. First of all, particle acceleration could produce a series of breakdowns of the particle involved. Depending on the number of

computations of data that you would want to make, this would depend on how many frequencies you would want to compute and by the substitution of one frequency for the other, the beat frequencies between the others may decrease or increase. Beat frequency is a disturbance between
uneven frequencies and by using selected frequencies for the purpose of computation, the beat frequency divided by the other frequencies added up, you could could equal values so small that they could literally be used as constant variables in calculus. For example, one frequency could be used as a divisor and the other frequencies added up as a sum could be the answer manifested as an average in beat frequency to yield division and this is just an example. Farther dissection of frequencies could be used for more complex mathematics. The acceleration of particles is also the acceleration of data.

This is why particle accelerators could be so great for super computation. The acceleration of particles is the physical analogy to the acceleration of data. The higher the speed is of particles, the more data that they are. The more rapid that the acceleration is, the faster the data can be collected and processed.

Signal Amplification by Harmonic Resonance

I was thinking that it might be possible to amplify a signal much more efficiently and with more crisp sound if a set of oscillators were equally modulated in their amplitude by an audio signal and were each at a harmonic multiple of a specific frequency.

These would be a set of Morley oscillators with the wire tapped in order from highest to lowest frequencies. The lowest frequency should be at the center and the higher frequencies should be placed in ascending order with respect to the turns in each winding..

The audio frequency input would according to my reasoning, be fed into each oscillator simultaneously. The harmonic multiples should be at multiples of 2x the frequency of the coil beneath.

The resonance between the harmonic frequencies would mathematically in the end produce one common frequency that could be electromagnetically induce a very high output voltage and/or frequency in a central coil with accordance to each number of turns
for each coil in the Morley oscillators. This is due to the concentrations of magnetic fields around the output coil.

The output signal of, course would have to be demodulated by silicon diodes. Two of these elaborations would be good for stereophonic amplification and output.
 Morley oscillators are oscillators that use amplification and feedback to produce a signal. An example of this feedback occurs when a connected microphone is held up to the output speaker.

The sensitivity and output of this amplifier are dependent on the ability of the coils to resonate when they interact with each other via the harmonic frequencies. The harmonic and initial frequencies should be in the radio frequencies.

Gravitational
Reflectivity

I have decided that Push Gravity (A friend named Richard told me about this type of gravity.) and the other way around are correct in theory. I think the edge of the universe is highly reflective because of possibility that space can act as a wall against exiting it. The reflectivity would have a field-like nature that would push inward and the other way
around, which I call relativistic gravity would act with the same force. The push would disappear tin this case. On the other hand, the relativistic gravity would disappear with the push gravity. Both would most likely be the same force. This would also allow for the universe to be in two simultaneous sates: 1. Singularity and 2. Zero-density.
This works very well with my theory that between infinite and zero density we have the a superfluid and this would be space-time. This would also allow for the difference between a nearly infinite temperature and absolute zero and the difference between matter and
energy, stationary and moving. This would also allow for a simultaneous expansion and collapse of the universe. Oh, yeah and the state of a quantum soup and that of super-
solidity.

This particular idea of mine seems to work well when Stephen Hawking's inflationary model of the universe is taken into consideration. For farther reference, I would consult the internet but it involves closed inflation, in which the universe has a surface tension and expands outward from the explosive force of the Big Bang singularity and from an
outside perspective while Open inflation is more from an inside
perspective in which a spatial surface tension is absent.

The Bermuda Triangle and Electromagnetic Vortexes

I have a thought that if electric and magnetic fields are overlapped, then this would be the same thing as looping electromagnetic fields in their unified form. If the electromagnetic field is looped, then maybe it might treat space and time as if space and time are paper and act as sewing thread in the same way that the paper is braided in a Chinese finger bracelet that you find at the arcade where you earn it for X number of tickets.This may explain the spatial disorientation that pilot feel over the Bermuda Triangle when they fly through similar or possibly same phenomena. This might possibly be recreated by

spinning a circular set of magnets around with a motor with each pole alternately placed in order to produce an electromagnetic vortex.

As a result of such an occurrence, space and time may twist and any traveling object may possibly move faster through twisted space-time followed by space and time twisting more tightly and vice versa.
Remember the Chinese Finger Bracelet?

This same effect may possibly, in my thoughts, be a way to achieve magnetic levitation if
the effect of an electromagnetic vortex can be used to bounce magnetic energy off of the ground and back up to the source or the magnetic vortex. Could this be the way that Adolf Hitler's electrical engineers designed the war machines discussed on The History
Channel? I hate Nazi mentalities and similar ideologies, and I don't encourage anyone to have the same ways of thinking either but it's interesting to know also the Volkswagon wasn't the only invention by otherwise sick and barbaric minds!

Breadth and Depth

If you want to contemplate belief in one topic with breadth you must think with depth and organize your thoughts coherently. If you want to contemplate a belief with depth then you must learn to manipulate the breadth in a coherent fashion which you organize it in a directed way. Organization of breadth is a composition made out of notes you take throughout your life and organize in an orderly fashion. This exists in theoretical physics and this exists in music. This is interesting to have developed as a concept because my older sister, Annie is a very good piano player as much as I am a physics freak and I never thought this would come to me, but music is a science and composition is the theoretical drive behind it. Annie's compositions and references and my compositions and references are a science and anything of a mastery is of the science of. Annie, I love you!

Particle Acceleration and Massive Breakdown

One thing I do notice about particle accelerators is that particles are might able to decompose during acceleration into their constituent frequencies and these frequencies each equal the energies of specific particles. Each particle at the speed of light might for instance decompose into photons or particles of light not by accumulation but actual decomposition of matter into its more fluid form which would be pure energy.

I think If the energies can be measured separately as frequencies and by the number of times per nanosecond that they occur which when multiplied by the velocity of each particle would when 1 is divided by the answer, equal the rarity but not only this, the speed of light divided by the rarity should equal the actual energy of these particles. at such speeds.

 Because of ratio of energy over rarity, the smallest particles by energy and mass should occur more commonly and as a result, I don't think the universe was ever 100% solid as a singularity but started out as a super-fluid ball of energy that was infinitely hot but as the energy pushed outward, the complexity of position may have taken over and formed relative space and time while the matter that curved it had a complexity level that equaled the complexity of the entire universe, so that it would be nearly impossible to break the mass of a body of matter up into the essential components unless there is a field-like formation which the progression of breakdown would form pure energy if all of it was ever broken down to its essential constituent form. This of, course would probably be high energy electromagnetic radiation or in other words, light.

By the time that it could be broken down to the point of pure energy, an electromagnetic charge would probably according to my reasoning span out to equal its equivalent plot in space and this could equal a lot of space. Imagine a particle smaller than an electron being able to produce an electromagnetic force stronger than the gravitational pull of a super-massive black hole.. Because of the architecture of fields described herein, the stronger the field between point A and point B, while moving in toward the body or particle of matter, the more greatly it represents the mass in an energetic state. The energetic aspect of a field is representative of the equivalent mass involved.

Last Theory, Big
Thanks.

What is time?

To end my book, I must comment that my first word was light. Also, I believed that the light must have come from something. When I got older, everything came from something else and I wondered what everything was made out of. When I saw the clock, I wondered what time was and if I could count and end at some number or if I could count for eternity. I have 1 answer to where everything ends and it isn't when, but it is where, making space fairly linear in concept with comparison to time. When I was 3, I had to ask why I couldn't see the air. Only my grandfather (Edward Durham), a war veteran and attorney could answer my question, but could not explain the origin of the universe. Time is a good thing ask about the origin of. Space is the linear form of time, but it is the progression of everything that can progress in form that makes time. Quantum mechanics can only describe in frame, but at the smallest levels of space, there is time and time is the beginning of and the cessation of all things with the progression of what happens to them. When time ceases, the universe disappears. The acceleration of any mass beyond the speed of light may mean faster computations and a smaller universe to compute. In the end, we will know so much that we will no longer need technology to communicate and the world may never be spied on again. These are my thoughts and memories with comparison to all other prediction...maybe all of the bad drivers will have to give of the cell when minds make technology obsolete, as those with a concept of time can be at any place that they wish and if only we were nomads with telepathy! A far off horizon, unless you allow your mind to travel and do what makes the mind travel best.

I spent 16 years of my life with slight memory loss and schizophrenia to have a conclusion that makes me think highly of my goal: To get my memory back. Well, I regained my memory and put everything in order. I must mention my interest in electricity, besides that chemistry is only a portion of what everything is made out of. To know the universe, you must know what everything is made out of and nobody does.

I must think Barbara Fogg for all that I could disasseble in school and I must thank Carina Jørgensen for helping me in my quest to regain my memory. Thank you both for what you have done for me. I have to thank Chris Ackerman for best friendship.

The Theory of Dimensional Physics Unification

by Nathaniel David Durham
Montrose, CO

My conclusion after so many years is that time is no more than the forces of energy. Anything that occurs in any way that involves energy also acts in an inertial way that is relative to all changes in all surroundings. So, therefore,time is no more that relative change. So, gravity may be a result of the action of time upon all energy. This could result in the solidification of all energy,eventually, back into matter. So likewise, even electrical charges could be a result of matter in an accelerated state of relativity. So suppose that 2 like states occur. This would cause repulsion by the fact that once not relative, then there is a cessation of attraction and therefore, produce repulsion. Strong nuclear force could therefore, result from the pure action of time on matter in the most dense possible state before the effect of gravitational collapse could occur. So suppose a disturbance of balance between 2 canceling forces and then apply this to time,as well. Weak force could then, be the most active force in determining the nature of every other given force by the direction of imbalance. This would also explain very well why matter has the potential to become infinite energy as well. This would provide an explanation for all phenomena in the universe. So now, there is the force of balance and force of cancellation. This determines that there can now be 6 existing forces. 2 forces per dimension and then the 7th force: Connection. This would also serve to prove Brain Greene correct in his book, "The Elegant Universe," that there are indeed, 11 dimensions.

SEQUENTIALISM

The Relative Theory of Physical Mathematics and Time Sequencing

by Nathaniel David Durham

35 S. 5th St.
Montrose, CO
81401

Will and thought are like space and time and determination is like space-time. Put them together and inhibition bends both and forms memory and this forms wisdom. So, this would be the mental equivalent of 4-dimensional space. Perception is like matter and the greater that is it, the greater that all can be bent until solid. This would be just like all 4 forces of matter being unified. Now, form a brain wave and then think about the overlap of concept. This bends it into 2 more,making 10 and then with the constant change, 12 and then 14 and then an infinite set of dimensions. meaning infinite mental potential and infinite potential of matter as parallel forms. This means unified physics. The brain, basically could release infinite energy in mental form with sufficient desire to use the mind.

I posted it as my status for philosophical teaching purposes. Mental force may from from the logical set of sequences that from from nonexistence having a double negative value that forms a universe, a void, an impulsive force and time. Space is the effect that impulsive force has on space and against the impulsive collapse and thus, the void and matter from these double negations. Now, take into instance that gravity results from the relation between both and that gravity is a result of the liberation of impulsive force in the opposite direction, inverting the value of space and time, forming a wave, collapsing time, magnifying existence and therefore, greater energy at greater alternation from the inversion of double negatives. This would form every force in the book. 1 that synchronizes with the expansion of the universe repels, while the opposite makes attraction, explaining all forces that when concentrated, make both a relative force, which is of change in value, compared to none. Nothing gets anywhere without change. This is a theory that I will call Relative Physical Mathematics.

the relation is that everything has to cancel a force to function. Theoretically speaking, a field would have to travel faster than the wave that on vibration, it radiates. This is because the higher that the energy is for particles that are smaller than a photon, if they are anything close to being massless...then to have a momentum significant enough for any physical action and for time and space to bend or warp, then these tiny particles would have to move faster than light, including gravitons and neutrinos. This, based on the written latter. I based this theory on my early childhood, during a vacation with my father and 2 sisters to Utah to see my cousins. I asked myself a simple question..."What is beyond the nature of light and what stretches farther?" Though at the time, I did not know that light moves, but I knew that it had to do something to get to me. This was shortly before I was diagnosed with cerebral palsy and both, being in the year, 1983.

In the end of it all, the distance between things will always involve some kind of relative force, as much as any type of difference will. So, the repulsion between 2 or more elements of existence would most likely be the result of a concentration of all potential forces that constitute energy. Anything else that happens in this universe is a secondary result of all impulsive reactions and energy is a force of impulse and not the other way around. So, anything that happens in this universe is impulsively driven. Thinking is just as impulsively driven as any other force, as well.

The Time Sequencing Theory

The past is never completely gone, but a folded set of dimensions, sequenced so that space and time are connected. The future is where sequencing begins, of all spatial dimensions and the present is the way that space and time are sequenced from a stationary frame of reference. This, thus the progression of time is the relative curvature of space, by matter. by Albert Einstein in his theory of relativity. Predicted that matter curves space and this gives my theory some of its basis.

This is the basic equation for the latter sets of theories: $\Delta \int d + d^2 \mid d(d^2) - y + x/2 = z = 0$ ^, meaning "to the power of," Δ means change, $\int$ means integral or integration of the value of and d means infinitesimal value. Z is the variable of all mean.

Finally, what is an expanding universe from a big bang, but something that has been pushed beyond its own properties to cause a big bang to happen. Not only is non-existence capable of this, but so is matter and so is equally, energy. Time is also the span of which things can and over which things will happen that will and thus, there is space-time.

Now suppose that everything that could happen in this universe over the span of eternity was collapsed into a small instant. An instant would equal a point in space, so space and time would be equal. Now suppose that in the collapse of time, that this process equally reversed. The universe would re-expand, but much more rapidly with more complex things in a single spot and possibly multiplied by infinity.

This is the most basic explanation that I wished to give in this book for what could bend every force in such a way as to form a rainbow of every force and in turn, matter.

You were right, Einstein. Something has to be a common origin for every force. Every force is most likely caused by the impact that time has on any alteration and alteration, time.

There is a force that must drive every action and be responsible for all actions and I believe that all that is is a result of the simplest of all things: an action. So I call this force Active Force.

The Energy Dynamics of Time

Time is no more than the relative difference, that arises from a concentration of change over a distance through space, which if happens quickly, concentrates the same energy, at first, in that local area and then radiates it at a decelerating rate. The energy then, travels through time, folding it (curvature), into the form of space and thus, the the thermodynamic explanation here, that I am giving that any radiation from any object is therefore, a form of evaporation and therefore, a form of cooling. The other effect is the mechanical slowing of what occurs. Thermal cooling occurs over a space and time, through distance by the radiation of thermal energy, atom to atom, but depending on atomic density, mass and conductivity of heat and electricity. Heat is the result of absorbed energy, trapped within the electrical relationship of the electrons and the nucleus, both being oppositely charged, thus why atoms must vibrate to radiate the heat into other atoms, through impact. However, in my experimental thought here, both must travel over a space and through time in the same way as in the previous latter. So, space and time are electrodynamic and thermodynamic and both heat and electricity are a similar force, but 1 can only be carried by vibration whereas the other can travel both ways, but in the end, all matter cools, once energy is lost. My explanation here, is that all energy is the same in form, but can only interact in the same way that that the medium that carries it can radiate it. So, this applies well, in chemistry where ionic bonds are electrical in form and covalent bonds are thermal in form. Together, this forms, hopefully, a pure dynamics of energy. Gravity is the result of the overlap of space and time, giving explanation that gravity produces higher temperatures in the same way as blue-shift of light occurs when a massive body is approached,redshift happening otherwise since matter acts to displace time, as well as space. The less time that there is, present, then the faster that energetic reactions occur. Strong nuclear force, which holds atoms together, is the concentration of all principles in the latter, where weak force would be a point where these principles act to the point that the given, supposed mass can no longer hold its composition, against the previously mentioned force of super-symmetry. Then, this is the point at which radioactive atoms are possible, beyond a specific atomic weight. Atoms, therefore, are a result of very powerful forces, acting against time in such a way that at absolute zero, they would not just disintegrate and cease to exist, not from a cold state, but as a result of radiating energy in the first place and so, absolute zero must occur as a result of the decay of matter into pure energy. Absolute zero may only happen at smaller levels of space than that of an atom, but it is possible. So, now at least I have a way to explain every force that I have studied.

The Unified Principle Theory

That it takes something for something to exist and that it takes something to exist and that it must be made out of something, is inertia. It is also, relativity in rational expression and minus its magnitude of existence, thus inertia. However, the universe could be just a novel, written to us to read and the higher being might be just a man in a novel, reading a novel by an even higher being, in in from, this is a relative and rational, proportional by limitation, mathematical expression. The end of existence is at the point of eternity and this is just the beginning of existence, making existence small, until eternity. The relation between these novels would compose the relative universe, while the other way is a quantum expression of the universe as a whole, but only expressible by relation. But then the novel would become the reader and the reader, the author. Any driven expression of energy could be a thought and all thoughts the same principle. Every impulse is a memory and everything that exists is the future, bent over the relative past, with the past at the center and the future as a force, as an orbiting curve, making a field around the dense past. This can make and electron out of our thin future and nuclei out of the concentrated and therefore, dense past, with opposite times producing the spin and charge of each particle. The relative makeup of the present nucleus would make up the neutral instant of relative present and thus, the neutron, which, due to its place in space and time, gives is a short wavelength to compensate for its lower mass than the proton and forming the total atomic mass when the electron is included. So, when an atom splits, time is split and the neutron is thus, released and this is called fission. When 2 nuclei join, time is displaced and collapses, releasing much more energy from the heightened vibration of each particle that makes the atom and this shortens wavelength, thus releasing more energy than if time had no influence. So, 2 present neutrons split each other into 2 times, thus repelling the neutron in a radioactive atom, which can contain either, 2 many neutrons, too many protons (repelled by same charge.). Then electrons can be emitted after a lack of protons, when protons are in excess in quantity. All squeezed into 1 as a principle, this makes an existence where infinitesimal value (equal to the value in calculus, d) to any variable of change, but multiplied by the mass and energy as well, all amounting, eventually to its potential at the point that eternity, where existence will cease. So, as humans as well, as people with disabilities, never doubt you potential and as an atheist, the novel comparison was to make an analogy and solve a paradox. The expression of this principle is equal to $\int(\int d\text{-} d \mid d) + d) \text{-}1 = 0$ and so, d gains significance here,, nearly a number compared to infinity. Time is the proportional relationship between all things, expressed with the same mathematical equation.

Never doubt yourselves in any way, because these variables can express potential in you, as you drive yourself to learn and accomplish new things.

www.ingramcontent.com/pod-product-compliance
Lightning Source LLC
Chambersburg PA
CBHW021906170526
45157CB00005B/1991